I0493871

NIST TECHNICAL NOTE 1711

Database-Assisted Design for Wind: Multiple Points-in-Time Approach

DongHun Yeo

Engineering Laboratory
National Institute of Standards and Technology
Gaithersburg, MD 20899-8611

February 2011

U.S. Department of Commerce
Dr. Gary Locke, *Secretary*

National Institute of Standards and Technology
Dr. Patrick D. Gallagher, *Director*

Disclaimers

(1) The policy of the NIST is to use the International System of Units in its technical communications. In this document however, works of authors outside NIST are cited which describe measurements in certain non-SI units. Thus, it is more practical to include the non-SI unit measurements from these references.

(2) Certain trade names or company products or procedures may be mentioned in the text to specify adequately the experimental procedure or equipment used. In no case does such identification imply recommendation or endorsement by the National Institute of Standards and Technology, nor does it imply that the products or procedures are the best available for the purpose.

Abstract

One of the problems encountered in the estimation of wind effects on high-rise structures is the development of combinations of such effects as translational responses and the rotational response to wind, and/or forces and moments at various cross sections of individual structural members. In current wind engineering practice such combinations are developed largely "by eye" since phase information on the effects being combined is not readily available from frequency domain analyses. In contrast, full time series analyses can produce estimates of combined wind effects, since they preserve phase information; however, such analyses can be too time-consuming.

To solve the problem, this study developed a time domain, database-assisted design (DAD) procedure that uses a multiple points-in-time (MPIT) approach, and illustrated the procedure by an application to a 60-story reinforced concrete structure. Results showed that the MPIT approach produces remarkably accurate estimates of the peak combined wind effects by using a limited number of peaks of the time histories of the individual wind effects being combined. Those estimates are obtained far more economically in terms of computational time than estimates based on conventional time domain analyses that use full time histories. It is noted that frequency domain techniques are not capable of performing accurate estimates of peak combined wind effects owing to the loss of phase information between the random processes being combined.

Keywords: Database-Assisted Design (DAD), mean recurrence interval, reinforced concrete, time-domain analysis, point-in-time approach, wind effects.

Acknowledgements

The author would like to thank Dr. Emil Simiu and Dr. Therese P. McAllister for valuable advice and comments. The wind tunnel data developed at the CRIACIV-DIC Boundary Layer Wind Tunnel, Prato, Italy were kindly provided by Dr. Ilaria Venanzi of the University of Perugia.

Contents

List of Figures

List of Tables

1. Introduction

One of the problems encountered in the estimation of wind effects on high-rise structures is the development of combinations of effects due to each of the translational responses and to the rotational response. In addition, the member design of such structures requires the estimation of the combined wind effects on individual internal force present in typical interaction equations used for member checking and design. Therefore, the combination of individual effects should be accurately estimated in the structural design for wind.

Current wind engineering practice is based largely on frequency domain techniques that entail the loss of phase information. Although the spectral densities and cross-spectra of various types of responses (e.g., axial force due to one modal translational response and axial force due to a second modal translational response) can be estimated individually, the estimation of the combination of those responses is not performed in accordance with physically rigorous models. Rather, a large number – as many as tens -- of wind effect combinations are considered that are believed to result in reasonably safe designs.

In contrast, time domain techniques preserve phase relationships among all the effects that come into play in structural design. In recent years the application of such techniques for estimating wind effects has become possible owing to progress in pressure measurement technology and the availability of improved computing capabilities.

The approach known as database-assisted design (DAD) uses calculations of time series of combined wind effects on individual members and for the assessment of serviceability. The DAD approach is applicable to both rigid and flexible buildings and has been introduced in Chapter C31 of the ASCE 7-10 Standard (ASCE 7-10, ASCE 2010) Commentary (Main and Fritz 2006; Yeo and Simiu in press). DAD uses time histories of wind tunnel pressures measured at a large number of ports to compute time series of wind effects for structural members, interstory drift ratios, accelerations, and identifies peaks of these time series.

Because hundreds of dynamic analyses need to be performed for various wind directions and wind speeds, the estimation of peak wind effects for all structural members of a high-rise building by conventional methods would require considerable computation times. A multiple points-in-time (MPIT) approach was therefore developed that provides an efficient and accurate estimate of peaks from combination of stationary time series of individual wind effects. The purpose of this report is to present the MPIT approach and demonstrate its application to a flexible reinforced concrete building (Yeo 2010; Yeo and Simiu in press). The approach can also be applied to other types of flexible buildings and to rigid buildings as well.

The report is organized as follows. Section 2 reviews expressions used in DAD for computing demand-to-capacity indexes (DCIs), interstory drift ratios, and top-floor acceleration. Section 3 describes the MPIT approach and uses a simple example to illustrate the procedure. Section 4 presents an application of the MPIT approach to a 60-story reinforced concrete building.

1.1 Multiple points-in-time (MPIT) approach

In engineering practice it is necessary to estimate the peak (i.e., the extreme value) of a combined effect resulting from two or more individual effects. Several approaches, including Turkstra's rule (Turkstra and Madsen 1980), the Ferry-Borges model (Ferry-Borges and Castanheta 1971), and Wen's load coincidence method (Wen 1977), have been developed to calculate peak effects. Turkstra's rule, which is empirical, yields reasonably satisfactory approximate results, and has achieved a high degree of acceptance. The alternative approach presented in this report improves upon Turkstra's rule and produces similarly efficient but considerably more accurate estimates of peak combined effects. Like Turkstra's rule, this alternative approach entails no restrictions with respect to the marginal distributions of the time series.

To introduce MPIT the following example is considered. The combined effect being considered is $X(t) = X_1(t) + X_2(t) + X_3(t)$, that consists of the sum of three time histories where t denotes the time. Each effect time history has 200 time steps. The n largest peaks are selected for each time series. Figure 1 shows time histories of $X_1(t)$, $X_2(t)$, $X_3(t)$, and $X(t)$ where the highest peaks of individual and combined effects are identified by circles for $n = 3$. The peak combined effect is estimated from a total of 9 (i.e., 3 individual effects × n) peaks in 9 points where 3 peaks of each individual effect occur. The peak of the combined wind effects is selected as the largest of the 9 values consisting of the 9 combined effects $X(t_i)$ ($i = 1, 2, ..., 9$) identified by circles in Figure 1(d). The estimated peaks are 9.25 for $n = 1$, and 10.57 for all values $2 \leq n \leq 200$. This shows that, in this case, the MPIT approach estimates reliably and efficiently the peak of combined effect using $n = 2$ peaks per time series of individual effect, instead of the full time series ($n = 200$). The combined effect in the example consists of linear combination of time series; however, linearity is not required in this approach, as will be shown subsequently.

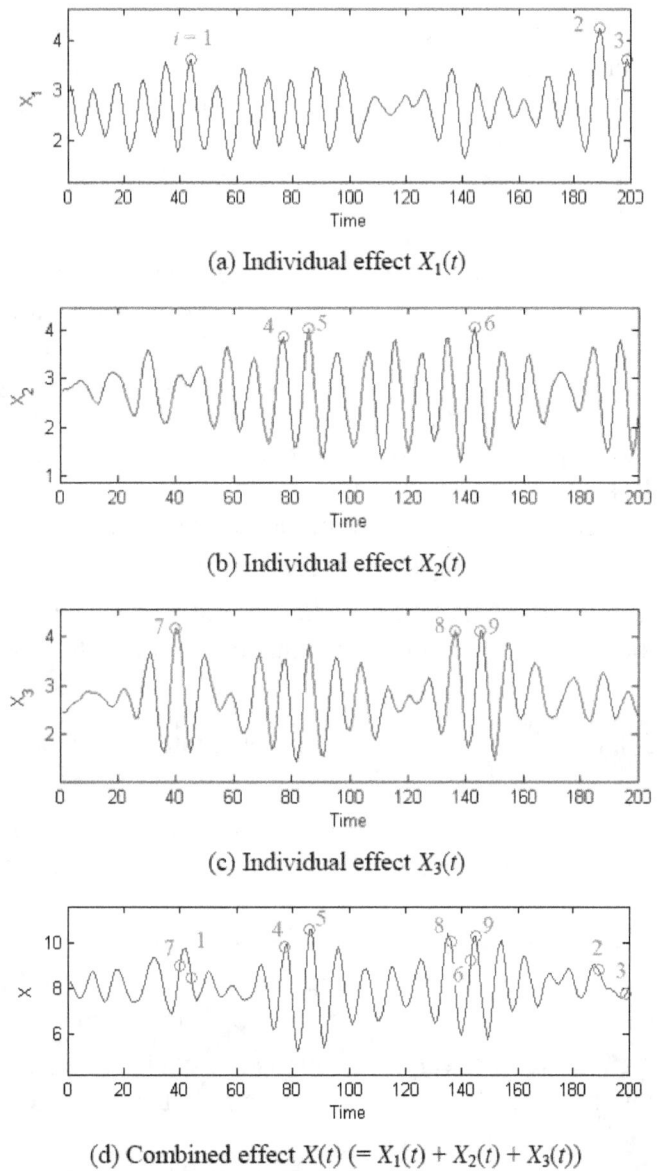

(a) Individual effect $X_1(t)$

(b) Individual effect $X_2(t)$

(c) Individual effect $X_3(t)$

(d) Combined effect $X(t)$ (= $X_1(t) + X_2(t) + X_3(t)$)

Figure 1. An example of MPIT approach

Note: Circles indicate the highest peaks within the time histories. There are n highest peaks in each individual time history, and $3n$ peaks in the combined effect at the corresponding ordinates of the time history of the peak individual effects.

11

2. Combined wind effects considered in design

The DAD procedure for the design of high-rise reinforced concrete buildings computes three types of time-history response for wind effects: (1) demand-to capacity indexes (DCIs) for forces and moments (Sect. 2.1), (2) interstory drift ratio (Sect. 2.2), and (3) acceleration at the top floor (Sect. 2.3).

2.1 *Demand-to-capacity index* (DCI)

For a member cross section, the DCI time series is defined as the ratio of internal forces and/or moments induced by the design loads divided by the corresponding design capacity. The member capacity is based on the Building Code Requirements for Structural Concrete and Commentary 318-08 (hereinafter ACI 318-08, ACI 2008). An index higher than unity indicates inadequate design; the index must be less or equal to unity for the design to be acceptable. DAD has two DCIs for combined loading: (1) axial and/or flexural loads, and (2) shear and torsional loads.

For cross sections subjected to axial force, P, and bending moments, M, the DCI is denoted by B_{ij}^{PM} (Yeo 2010; the subscript i indicates the member i and the subscript j indicates the cross section j of that member). In the case of beams and columns subjected to the interaction of axial force and bending moments, the DCI has the simple expression:

$$B_{ij}^{PM}(t) = \frac{M_u(t)}{\phi_m M_n} \qquad \text{(for tension-controlled sections)}$$
$$= \frac{P_u(t)}{\phi_p P_n} \qquad \text{(for compression-controlled sections)} \tag{1}$$

where $M_u(t)$ and $P_u(t)$ are time series of the design bending moment and the design axial force, respectively, at the cross section being considered, M_n and P_n are the nominal moment and axial strengths at the section, and ϕ_m and ϕ_p are reduction factors for flexural and axial strength, respectively. Equations in Eq. (1) are expanded in the following equations to show how P, M_x, and M_y are accounted for in tension and compression controlled sections of beams and columns.

For tension-controlled sections subject to bi-axial flexure loads, the PCA (Portland Cement Association) load contour method (PCA 2008) is used:

$$B_{ij}^{PM}(t) = \frac{M_{ux}(t)}{\phi M_{nox}}\left(\frac{1-\beta}{\beta}\right) + \frac{M_{uy}(t)}{\phi M_{noy}} \qquad \textit{for} \quad \frac{M_{uy}(t)}{M_{ux}(t)} > \frac{M_{noy}}{M_{nox}}$$
$$= \frac{M_{ux}(t)}{\phi M_{nox}} + \frac{M_{uy}(t)}{\phi M_{noy}}\left(\frac{1-\beta}{\beta}\right) \qquad \textit{for} \quad \frac{M_{uy}(t)}{M_{ux}(t)} < \frac{M_{noy}}{M_{nox}} \tag{2}$$

12

where M_{ux} (t) is the design bending moment about x-axis, M_{uy} is the design bending moment about y-axis, M_{nox} is the nominal uniaxial moment strength about x-axis, M_{noy} is the nominal uniaxial moment strength about y-axis, and β is a constant dependent upon the properties and details of the member, for which the value 0.65 is typically used as an approximation. The x and y axes are the principal axes of the cross-section under consideration. Note that M_{ux} and M_{uy} are dominant over P_u is small for this condition and therefore P_u is not included in Eq. (2)

For compression-controlled sections the Bresler reciprocal load method (ACI 318-08 (R10.3.6)) is used:

$$B_{ij}^{PM}(t) = \frac{P_u(t)}{\phi P_n}$$

$$= \frac{P_u(t)}{\phi \dfrac{1}{\dfrac{1}{P_{ox}} + \dfrac{1}{P_{oy}} - \dfrac{1}{P_o}}} \tag{3}$$

where P_{ox} is the maximum uniaxial load strength of the column with moment $M_{nx} = P_n e_y$ (e_y is the eccentricity along y-axis), P_{oy} is the maximum uniaxial load strength of the column with moment $M_{ny} = P_n e_x$ (e_x is the eccentricity along x-axis), and P_o is the maximum axial load strength with no applied moments.

The DCI of a member for sections subjected to shear forces and torsional moment is denoted by B_{ij}^{VT}:

$$B_{ij}^{VT}(t) = \frac{\sqrt{[V_u(t)]^2 + \left(\dfrac{T_u(t)p_h b_w d}{1.7 A_{oh}^2}\right)^2}}{\phi_v (V_c + V_s)} \tag{4}$$

where V_c and V_s are the nominal shear strengths provided by concrete and by reinforcement, respectively, $V_u(t)$ is time series of the shear force, $T_u(t)$ is time series of the torsional moment, ϕ_v is the reduction factors for shear strengths, p_h is the perimeter enclosed by the centerline of the outermost closed stirrups, A_{oh} is the area enclosed by the centerline of the outermost closed stirrups, b_w is the width of the member, and d is the distance from extreme compression fiber to the centroid of longitudinal tension reinforcement. Note that V_c can be reduced by a tensile force $P_u(t)$ on the section (see Section 11.2 in ACI 318-08).

For sections subject to bi-axial shear forces the index is

$$B_{ij}^{VT}(t) = \frac{\sqrt{[V_{ux}(t)]^2 + [V_{uy}(t)]^2 + \left(\dfrac{T_u(t)p_h b_w d}{1.7 A_{oh}^2}\right)^2}}{\phi_v (V_c + V_s)} \tag{5}$$

13

where $V_{ux}(t)$ and $V_{uy}(t)$ are time series of the shear forces in the x and y axes, respectively.

2.2 Interstory drift ratio

The time series of the interstory drift ratio at the ith story, $d_{i,x}(t)$ and $d_{i,y}(t)$, corresponding to the x and y axes, are:

$$d_{i,x}(t) = \frac{\left[x_i(t) - D_{i,y}\,\theta_i(t)\right] - \left[x_{i-1}(t) - D_{i-1,y}\,\theta_{i-1}(t)\right]}{h_i}$$

$$d_{i,y}(t) = \frac{\left[y_i(t) + D_{i,x}\,\theta_i(t)\right] - \left[y_{i-1}(t) + D_{i-1,x}\,\theta_{i-1}(t)\right]}{h_i}$$

(6)

where $x_i(t)$, $y_i(t)$, and $\theta_i(t)$ are the displacements and rotation at the mass center at the ith floor, $D_{i,x}$ and $D_{i,y}$ are distances along the x and y axes from the mass center on the ith floor to the point of interest on that floor (Figure 2), and h_i is the ith story height between mass centers of the ith and the $i\text{-}1^{th}$ floor.

The ASCE 7-10 Commentary for Appendix C (CC.1.2) suggests limits on the order of 1/600 to 1/400 for those ratios.

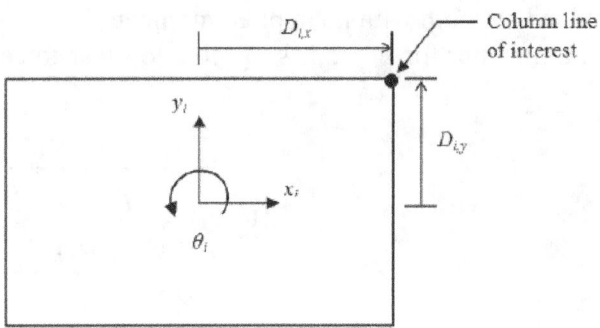

Figure 2. Position parameters at floor i for interstory drift ratio

2.3 Top floor acceleration

The time series of resultant acceleration at the top floor, $a_r(t)$ are computed in DAD by the expression:

$$a_r(t) = \sqrt{\left[\ddot{x}_{top}(t) - D_{top,y}\,\ddot{\theta}_{top}(t)\right]^2 + \left[\ddot{y}_{top}(t) + D_{top,x}\,\ddot{\theta}_{top}(t)\right]^2}$$

(7)

where accelerations $\ddot{x}_{top}(t)$, $\ddot{y}_{top}(t)$, and $\ddot{\theta}_{top}(t)$ of the mass center at the top floor pertain to the x, y, and θ (i.e., rotational) axes, and $D_{top,x}$ and $D_{top,y}$ are the distances along the x and y axes from the mass center to the point of interest on the top floor.

14

The resultant value of Eq. (7) is used, rather than accelerations along the principal axes, because peak acceleration is of concern for human discomfort. While ASCE 7-10 does not provide wind-related peak acceleration limits, for office buildings a limit of 25 mg with a 10-year mean return interval (MRI) was suggested by Isyumov et al. (1992) and Kareem et al. (1999). Note that mg denotes 10E-3 g, where g is the gravitational acceleration.

3. Application of MPIT approach to wind effects

Construction of response databases (or wind effect databases) requires considerable computational time. In particular, this is the case when calculating the DCIs for thousands of structural members in high-rise buildings. As shown in Section 2, DCI time series may comprise individual wind effects -- such as shear forces and torsion -- that interact non-linearly.

Response databases of combined wind effects (i.e., DCI, interstory drift ratio, and top-floor acceleration) provide the peak effects corresponding to wind speeds (e.g., 20 m/s, 30m/s, ..., 80 m/s) and wind directions (e.g., 0°, 10°, ..., 350°) of interest. They are obtained from their full time histories of aerodynamic pressures on a building envelope for all wind speeds and directions. Structural dynamic analyses are performed for each wind speed and direction, using time-series of aerodynamic forces acting at the mass center of each floor based on the aerodynamic pressures. For each wind speed and direction, the responses are obtained from the corresponding directional aerodynamic and inertial floor loads multiplied by the appropriate influence coefficients calculated by conventional structural analysis programs. The response databases consist of peaks of combined wind effects due to all possible wind.

Let the number of peaks selected for the MPIT approach be n. Then, n rank-ordered peaks for each time history of individual effects (i.e., internal forces and moments) are used for calculation of the peak DCIs. For the calculation of B_{ij}^{PM}, n highest negative peaks and n highest peaks are selected for the time series of P_u and n highest absolute peaks from the time histories of M_{ux} and M_{uy}. The positive and negative peak values of P_u are associated with axial tension and compression, respectively. For the calculation of B_{ij}^{VT}, n highest absolute peaks are selected from the time histories of V_{ux}, V_{uy}, and T_u. Additional n highest peak values of P_u are used for B_{ij}^{VT}, since the shear strength of a section is reduced by a significant tensile axial force (see Section 11.2 in ACI 318-08). Thus, in general there will be a total of $4n$ peak values to consider for each DCI. To establish a DCI response database in this case, $4n$ peak values are used for each wind direction and speed being considered. However, because the peak values can occur simultaneously (e.g., peaks of axial force and bending moment occur at the same time), in general less than $4n$ are actually needed.

For an example of the MPIT approach, consider the estimation of B_{ij}^{PM} of a column due to a specified wind speed and direction. Set the number of peak values for each time history of individual effects (P_u, M_{ux}, and M_{uy}) to be $n = 5$ for 10 peaks (i.e., the 5 highest peaks and the 5 lowest peaks) of P_u, 5 peaks from the time history of M_{ux} and M_{uy}, as shown in Figure 3, where the peaks are depicted as hollow circles. Thus, the total number of time points, n^*, at which the DCI is calculated is $4 \times 5 = 20$. However, some of those times coincide, so $n^* = 16$ of times are used for the B_{ij}^{PM} calculation in this example. The MPIT approach computes 16 DCI values per member, while a full time-series (FT) approach would calculate 7305 DCI values per member (i.e., one value for each of the ordinates of the time series in this example. The DCI plot (Fig. 3d) shows that the estimated

16

peak of MPIT is identical to the observed peak of the FT. Peaks of individual time series are not necessarily peaks in the resultant series. None of the resultant peaks corresponds to M_{ux}. This observation is indicative of the importance of phase information.

The application of the MPIT approach in this example is seen to result in reliable estimates of the highest peak of the combined wind effect and to reduce significantly calculation times. An effective MPIT approach requires the selection of a number n, which should be sufficiently small for computational efficiency and sufficiently large to yield accurate results.

The MPIT approach can typically be applied to the estimation of DCI response databases (i.e., DCI values as functions of wind speed and direction), which requires large computational times owing to the large number of members in a building. The MPIT approach need not be employed for the estimation of peak interstory drift and peak acceleration values, since the reduction of the associated computational time is small. For example, for the 60-story building studied in this section, each DCI requires 7800 sets of calculations (one for each of the 7800 members), whereas interstory drift along a column line requires only 60 sets of calculations (i.e., one for each story), and top-floor acceleration for a corner requires just one set of calculations. In addition, the calculation of interstory drift and accelerations is much simpler than DCI calculations.

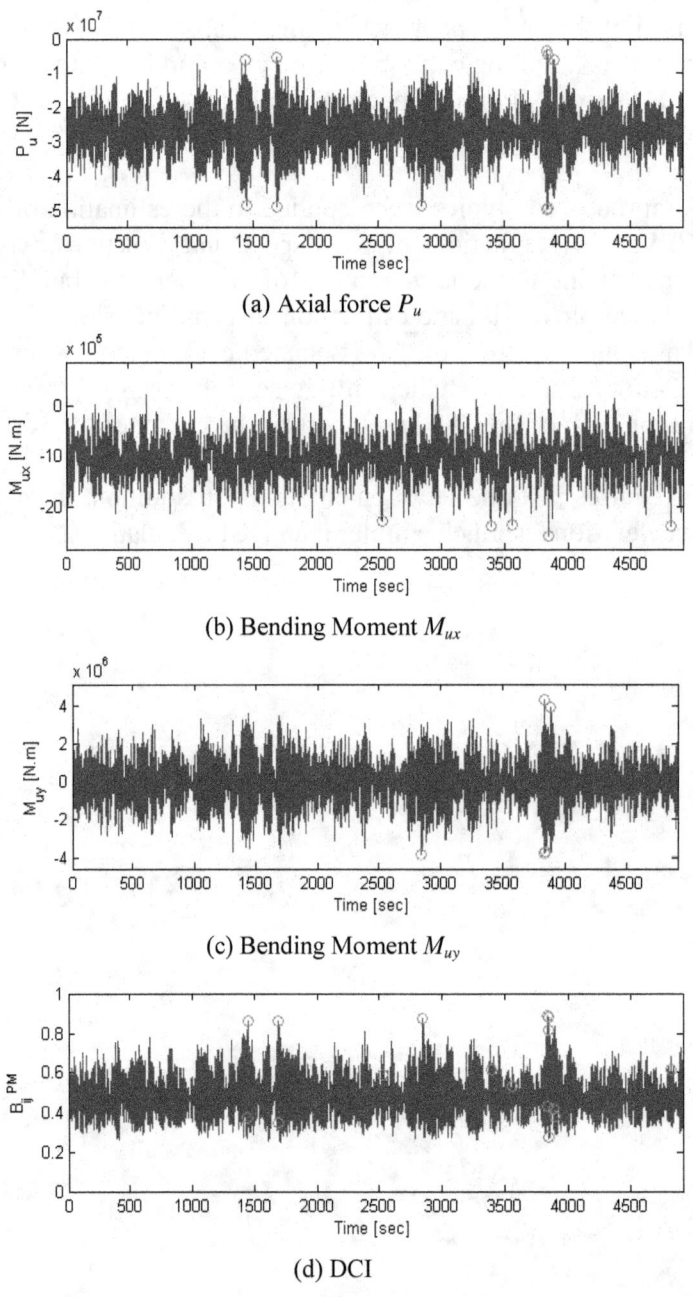

(a) Axial force P_u

(b) Bending Moment M_{ux}

(c) Bending Moment M_{uy}

(d) DCI

Figure 3. Peaks in time histories

4. Application to a 60-story CAARC building

A high-rise reinforced concrete building was evaluated using the High-Rise Database-Assisted Design for Reinforced Concrete structures (HR_DAD_RC version 1.0) software and the multiple points-in-time (MPIT) approach. The MPIT approach was based on the following numbers n of peaks: n = 1, 3, 5, 10, 20, and 40, for each structural member and its time series of forces and moments. The optimal number n was determined by comparing the results of the calculations from the MPIT approach to those calculated from the full-time (FT) approach..

The design building is a 60-story reinforced concrete building with rigid diaphragm floors (Figure 4) and is known as the Commonwealth Advisory Aeronautical Research Council (CAARC) building (Melbourne 1980; Venanzi 2005; Wardlaw and Moss 1971). The dimensions of the building are 45.72 m width (dimension B), 30.48 m depth (D), and 182.88 m height (H). The building has a moment-resisting frame structural system similar to the structural system, with comparable dimensions, studied by Teshigawara (2001), and consists of 7800 members (i.e., 2880 columns and 4920 beams). The building was assumed to be located in suburban terrain exposure near Miami, Florida.

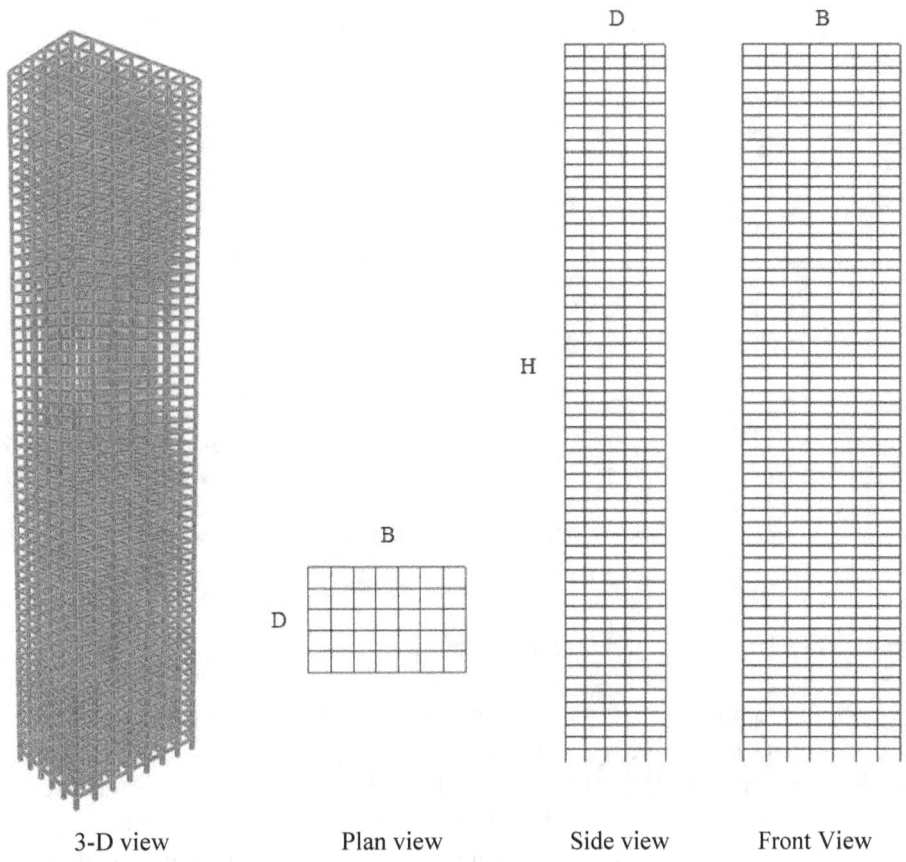

| 3-D view | Plan view | Side view | Front View |

Figure 4. Schematic views of a 60-story building

4.1 Modeling of the building

Structural members of the building consist of columns, beams, and slabs. Columns are divided into corner and non-corner columns, and beams are divided into exterior (spandrel) and interior beams. As shown in Table 1, the building is comprised of six sets of members for each member type. Each set consists of 10 stories where the member dimensions and reinforcement details are the same. The first set applies to the first ten stories, the second to the next ten stories, and so forth. The compressive strengths of concrete for all members are 80 MPa from the first to the 40th story and 60 MPa from the 41st to the 60th story. Columns have longitudinal reinforcement uniformly distributed along the sides and hoops, and beams have tensile and compression reinforcement and stirrups. The yield strengths of reinforcements are 520 MPa for longitudinal bars and 420 MPa for hoop or stirrup bars. In this study, wind effects were calculated for a typical set of 96 members out of 7800 beams and columns and slabs were not considered.

For dynamic properties of the design building, natural frequencies of vibration considered in this study are 0.165 Hz for the 1st mode in the y direction, 0.175 Hz for the 2nd mode in the x direction, and 0.200 Hz for the θ direction, calculated from a separate modal analysis (Figure 5). The corresponding modal damping ratios were assumed to be 2 % in all three modes.

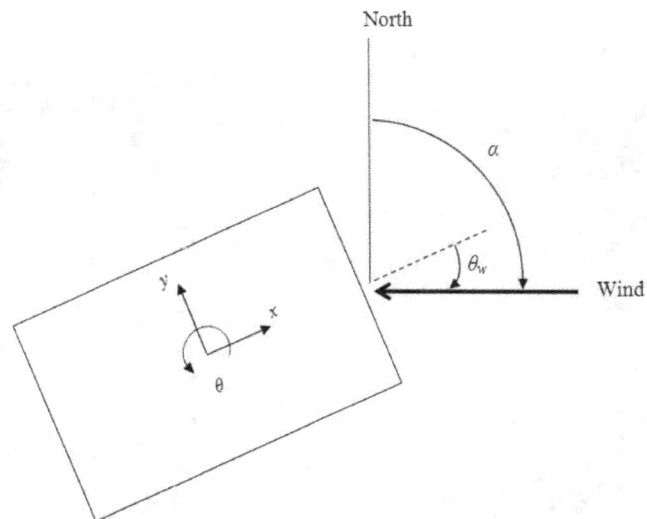

Figure 5. Local coordinates of the building and wind directions

4.2 Response database from aerodynamic pressure data

For wind with speeds of 20 m/s to 80 m/s in increments of 10 m/s and wind directions of 0° to 350° in increments of 10°, dynamic analyses were performed using time histories of aerodynamic wind loads at the mass center of each floor, calculated from time-series of aerodynamic pressures on a rigid model of the CAARC building measured in wind tunnel tests by Venanzi (2005). The analyses yielded time series of motion (displacements and accelerations) and effective lateral wind loads at the mass centers. The motion time series yielded values of interstory drift and the top floor acceleration. The lateral loads due to

20

wind, multiplied by influence coefficients, yielded internal forces and moments at critical sections of members. The combination of wind-induced internal forces and moments with internal forces and moments due to gravity load, using the design load combinations specified by ASCE 7-10, Section 2.3, yielded combined DCIs at the critical cross sections. This study accounts for two load combination cases (LC1 and LC2) for strength design, and one case (LC3) for serviceability design:

$$1.2D + 1.0L + 1.0W \qquad \text{(LC1)}$$
$$0.9D + 1.0W \qquad \text{(LC2)} \qquad (8)$$
$$1.0D + 1.0L + 1.0W \qquad \text{(LC3)}$$

where D is the total dead load, L is the live load, W is the wind load. The load factor of W for serviceability checks (LC3) can be reduced to 0.5 by designers (ASCE 7-10 Commentary, Section CC.1.2, Eq. CC-3). Note that the wind load in HR_DAD_RC is not factored but based on specified MRIs appropriate for strength and serviceability designs.

Response databases for DCI, interstory drift, and acceleration were constructed using peak DCIs for each wind direction and each wind speed. Thus, once a wind direction and a wind speed are specified, the associated combined wind effects can be obtained using the response databases. The construction of response databases requires a considerable amount of computation, since hundreds of dynamic analyses need to be performed for the various wind directions and wind speeds. Response databases of DCIs may be required for all members, which entails a large amount of computational time for up to thousands of members. The MPIT approach can significantly reduce the time when the response databases are constructed.

Figure 7, in which θ_w denotes the wind direction, shows the response database of a DCI for a corner column (cc1) under load combination LC1.

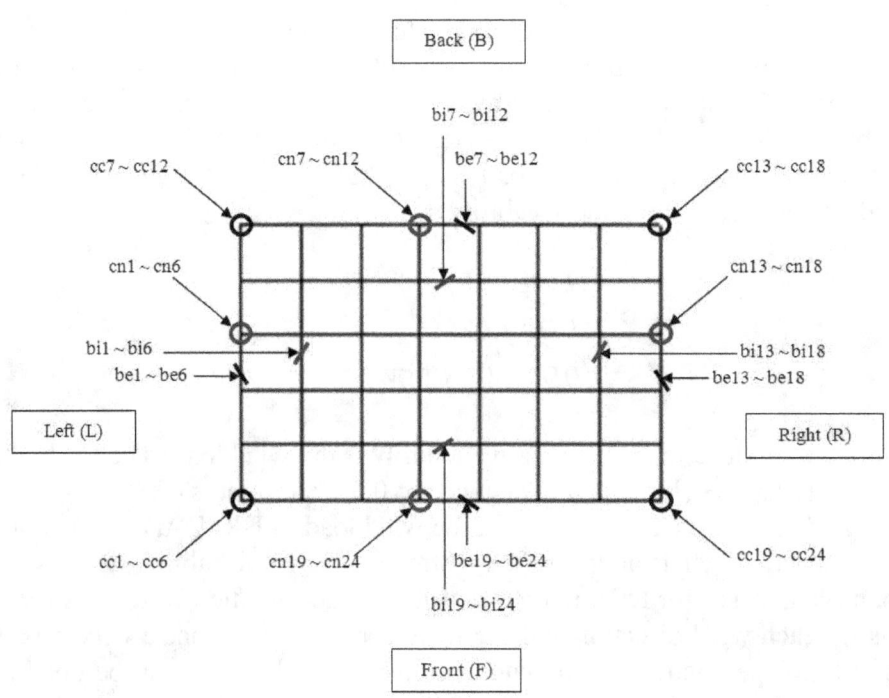

Figure 6. Plan view of building with locations of selected members ($\alpha_0 = 0°$)
(cc = corner column; cn = non-corner column; be = exterior beam; bi = interior beam)

Table 1. Section dimensions and reinforcement details for critical sections of structural members

Name	Story	Section [mm × mm]	Longitudinal bar	Hoop or stirrup [spacing: mm]	Selected member
Corner column (cc)	51st~60th	750 × 750	12 - D29	4 - D13@200	6, 12, 18, 24 (51st st.)
	41st~50th	750 × 750	12 - D29	4 - D13@200	5, 11, 17, 23 (41st st.)
	31st~40th	800 × 800	16 - D32	4 - D13@200	4, 10, 16, 22 (31st st.)
	21st~30th	850 × 850	20 - D32	4 - D16@200	3, 9, 15, 21 (21st st.)
	11th~20th	900 × 900	20+12 - D43	4 - D16@200	2, 8, 14, 20 (11th st.)
	1st~10th	1100 × 1100	24+16 - D43	4 - D16@200	1, 7, 13, 19 (1st st.)
Non-corner column (cn)	51st~60th	750 × 750	12 - D25	4 - D13@200	6, 12, 18, 24 (51st st.)
	41st~50th	750 × 750	12 - D25	4 - D13@200	5, 11, 17, 23 (41st st.)
	31st~40th	800 × 800	12 - D25	4 - D16@200	4, 10, 16, 22 (31st st.)
	21st~30th	850 × 850	16 - D29	4 - D16@200	3, 9, 15, 21 (21st st.)
	11th~20th	900 × 900	20+12 - D43	4 - D16@200	2, 8, 14, 20 (11th st.)
	1st~10th	1100 × 1100	20+16 - D43	4 - D16@200	1, 7, 13, 19 (1st st.)
Exterior beam (be)	51st~60th	400 × 700	4 - D32 / 2 - D32	2 - D13@150	6, 12, 18, 24 (roof)
	41st~50th	400 × 700	4+4 - D32 / 3 - D32	2 - D16@150	5, 11, 17, 23 (50th fl.)
	31st~40th	450 × 750	4+4 - D36 / 4 - D32	4 - D16@150	4, 10, 16, 22 (40th fl.)
	21st~30th	500 × 750	5+5 - D36 / 4 - D36	4 - D16@150	3, 9, 15, 21 (30th fl.)
	11th~20th	550 × 750	5+5 - D43 / 4 - D36	4 - D16@150	2, 8, 14, 20 (20th fl.)
	1st~10th	550 × 800	5+5 - D43 / 4 - D36	4 - D16@150	1, 7, 13, 19 (10th fl.)
Interior beam (bi)	51st~60th	400 × 700	4 - D29 / 2 - D29	2 - D13@150	6, 12, 18, 24 (roof)
	41st~50th	400 × 700	4+4 - D32 / 2 - D32	2 - D13@150	5, 11, 17, 23 (50th fl.)
	31st~40th	450 × 750	4+4 - D36 / 3 - D32	4 - D13@150	4, 10, 16, 22 (40th fl.)
	21st~30th	500 × 750	5+5 - D36 / 4 - D36	4 - D13@150	3, 9, 15, 21 (30th fl.)
	11th~20th	550 × 750	5+5 - D36 / 4 - D36	4 - D13@150	2, 8, 14, 20 (20th fl.)
	1st~10th	550 × 800	5+5 - D36 / 4 - D36	4 - D13@150	1, 7, 13, 19 (10th fl.)

Note: cc1~cc24 for corner columns; cn1~cn24 for non-corner columns; be1~be24 for exterior beams; bi1~bi24 for interior beams; st. for story; fl. for floor.

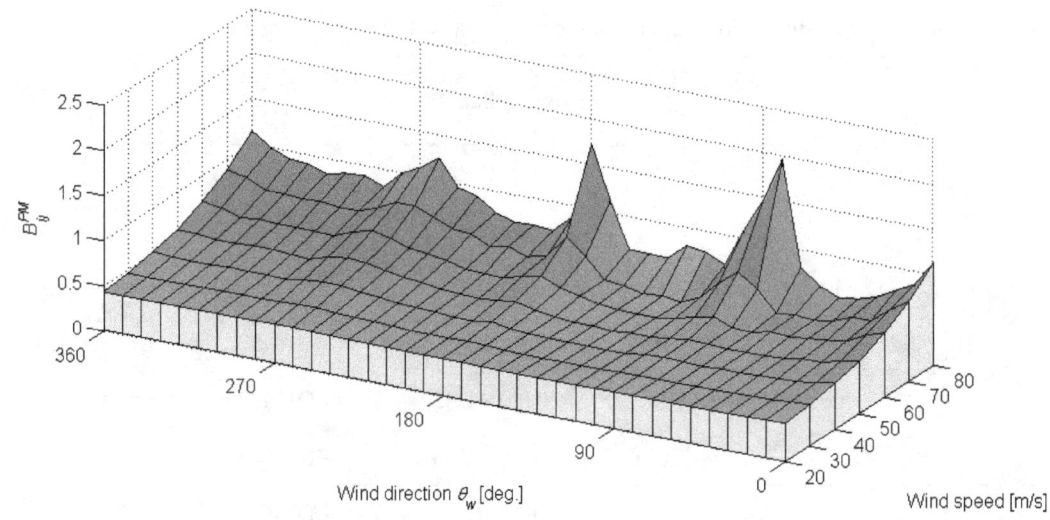

Figure 7. Response database: DCI (member ID = cc1)

4.3 Structural responses due to wind climate

Structural responses for a wind climate near Miami were obtained by applying the directional wind speeds from the climatological database to the response databases. The climatological database is a dataset of 999 simulated hurricanes with wind speeds for 16 directions near Miami, Florida (Milepost 1450, available at www.nist.gov/wind). The left side of the building was assumed to face South (i.e., $\alpha = 0°$ in Figure 5).

The terrain exposure near the building was assumed to be suburban (i.e., Exposure Category B) in all directions. The DAD procedure modified the climatological database of directional wind speeds by converting them to hourly mean wind speeds (m/s) at the building rooftop in suburban terrain exposure (see Section 11.2 in Simiu 2011, and Section 26.9.5 in ASCE 7-10). The climatological database was then applied to each response database to obtain the 999 peak responses. The peak responses corresponding to specified MRIs of the wind effects were estimated using nonparametric methods described in Section 12.7 of Simiu (2011).

Plots of peak response databases for LC1 are shown for DCIs of the corner column cc1 (Figure 8), interstory drift of the front-left corner at the 43rd story (Figure 9a), and peak accelerations of the front-left corner of the top floor (Figure 9b). Note that the peak interstory drifts and accelerations along the two principal axes of the building do not occur at the same time.

24

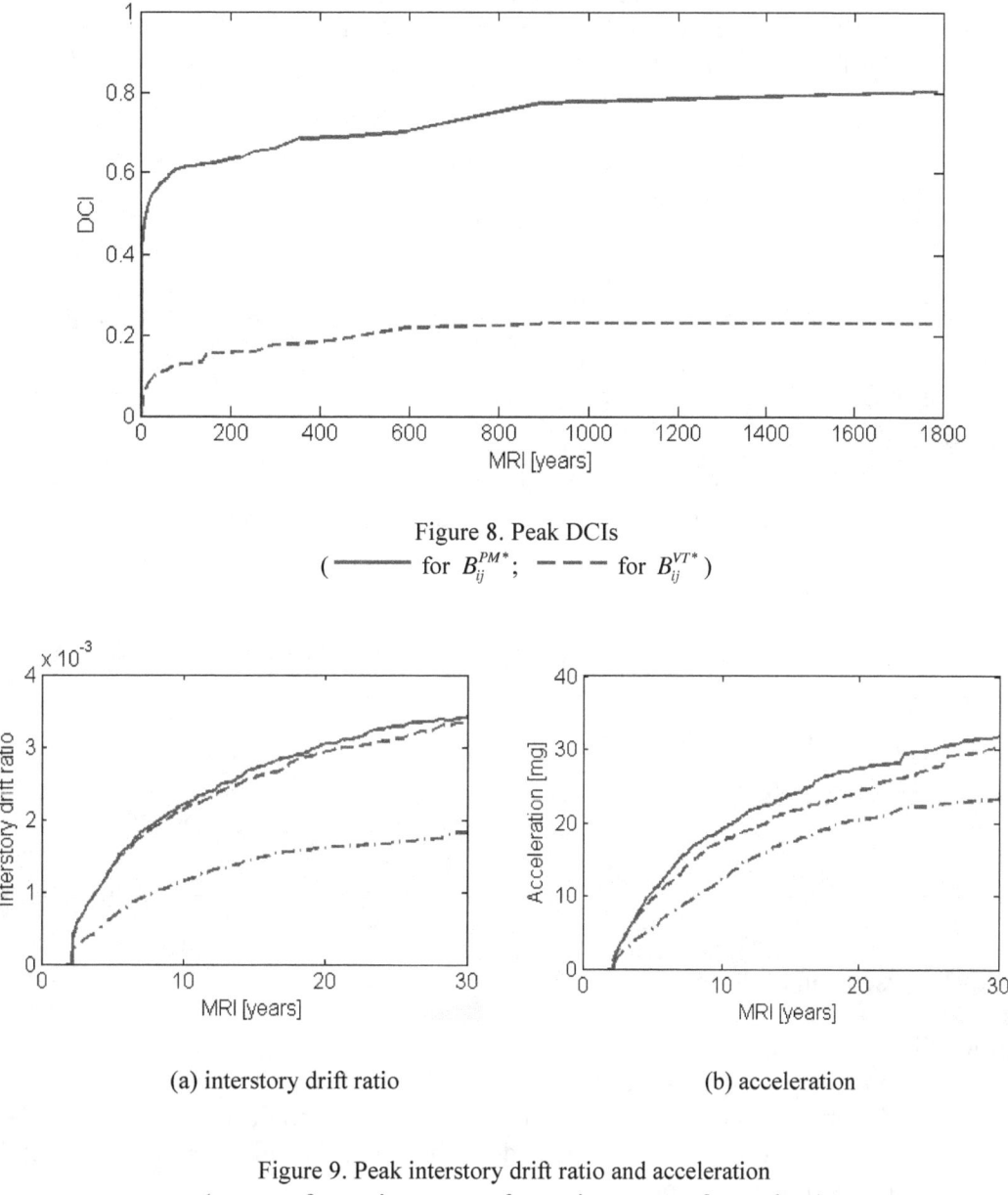

Figure 8. Peak DCIs
(———— for B_{ij}^{PM*}; – – – for B_{ij}^{VT*})

(a) interstory drift ratio

(b) acceleration

Figure 9. Peak interstory drift ratio and acceleration
(– · – · · for x axis;   – – – for y axis;   ———— for resultant)

4.4 Adjustment of demand-to-capacity indexes

ASCE 7-10 requires that overturning moments determined by wind tunnel testing must not be less than 80 percent of their ASCE 7-based counterparts (see ASCE 7-10, Section 31.4.3). Since DAD is based on wind tunnel data, this requirement applies to the computations in DAD. ASCE 7-based overturning moments about the principal axes (i.e., x and y axes) of buildings with Risk Category III and IV were therefore calculated for a basic wind speed of 81 m/s based on MRI = 1700 years (Table C26.5-3 in ASCE 7-10), and were compared to the peak overturning moments determined by the DAD for that MRI.

If the overturning moments in DAD are less than 80 percent of those determined in accordance with Part I of Chapter 27 of ASCE 7-10, the DCIs were adjusted as follows:

$$B_{ij}^* = \gamma B_{ij}$$

$$\gamma = \frac{0.8}{M_o^{DAD}/M_o^{ASCE7}} \tag{9}$$

where M_o^{DAD} and M_o^{ASCE7} are the overturning moments obtained from DAD and Part I of Chapter 27, ASCE 7-10, respectively, and γ is the index adjustment factor. If the moment in DAD is not less than 80 percent of the ASCE 7-10 value, the index need not be modified (i.e., $B_{ij}^* = B_{ij}$).

As shown in Table 2, ratios of overturning moments from DAD to those from ASCE 7 are less than 0.8 in the x axis and the corresponding index adjustment factor γ (Eq. (9)) is 1.12. Adjusted DCIs for MRI = 1700 years were obtained by multiplying the indexes by the adjustment factors.

Table 2. Overturning moments and adjustment factor

	M_{ox} [$\times 10^6$ kN·m]	M_{oy} [$\times 10^6$ kN·m]	$M_{ox}^{DAD}/M_{ox}^{ASCE7}$	$M_{oy}^{DAD}/M_{oy}^{ASCE7}$	γ
ASCE 7-10	6.49	3.92			
			0.72	0.97	1.12
DAD	4.64	3.81			

4.5   MPIT-based wind effects

Both MPIT and FT approaches were used to calculate peak wind effects (i.e., adjusted DCIs) corresponding to a 1700-year MRI. The DCIs of 96 selected members were computed for load combination cases LC1 and LC2. The MPIT approach employed numbers n equal to 1, 3, 5, 10, 20, and 40. The DCIs described subsequently are adjusted with the adjustment factors shown in Table 2.

Figure 10 shows DCIs of the corner column cc1 computed by both approaches. As n increases, the estimated peak value of a limited number of DCIs in MPIT rapidly converges to the observed peak value of full series of DCIs in FT. Table 3 compares DCIs for all 96 members calculated by using the MPIT and FT approaches. N_m denotes the number of members (out of 96) whose DCIs based on MPIT are not identical to the values based on FT, and R_m denotes the lowest ratio of MPIT-based DCI to FT-based DCI. The MPIT approach calculated the DCI for at most $4n$ local peak points (e.g. 12 points for $n = 3$, see Section 3). In contrast, the FT approach used 7305 points of the full time history. As shown by the results of the calculations, both DCIs estimated from MPIT using $n \geq 3$ are at least 98 % of DCI calculated from FT. This shows that the MPIT approach is an efficient and reliable method for calculating peak DCIs for linear or non-linear combinations of individual wind effects.

Table 4 lists, for $n = 3$, MPIT-based DCIs consisting of the maximum DCI value for each load combination out of all 96 members. The highest $B_{ij}^{PM^*}$ is 0.88 and the highest $B_{ij}^{VT^*}$ is 0.62, meaning that structural members were adequately designed for strength. That is, all members have the capacity to resist effects of interacting axial force and bending moments as well as effects of interacting shear forces and torsional moment corresponding to an MRI of 1700 years.

Calculated peak interstory drift for all floors for an MRI of 20 years and peak acceleration on the top floor for an MRI of 10 years at a corner intersecting the front and the left sides of the building (see Figure 6) are summarized in Table 5. The table also lists peak values along the principal x- and y-axis, and the associated resultant for the interstory drift ratio at the 43rd story and the top floor accelerations. Note that the largest interstory drift occurs at the 43rd story. The estimated peak interstory drift ratio is 0.0029 in the y direction, which exceeds the 1/400 limit suggested by the ASCE 7-10 Commentary, Section CC.1.2). The estimated peak top floor resultant acceleration is 19.3 mg for an MRI of 10 years, which is lower than the 25 mg limit suggested by Isyumov et al. (1992), indicating that the design may be adequate for peak acceleration. As stated in Section 3, the MPIT approach was not applied to calculating the peak wind effects for serviceability design.

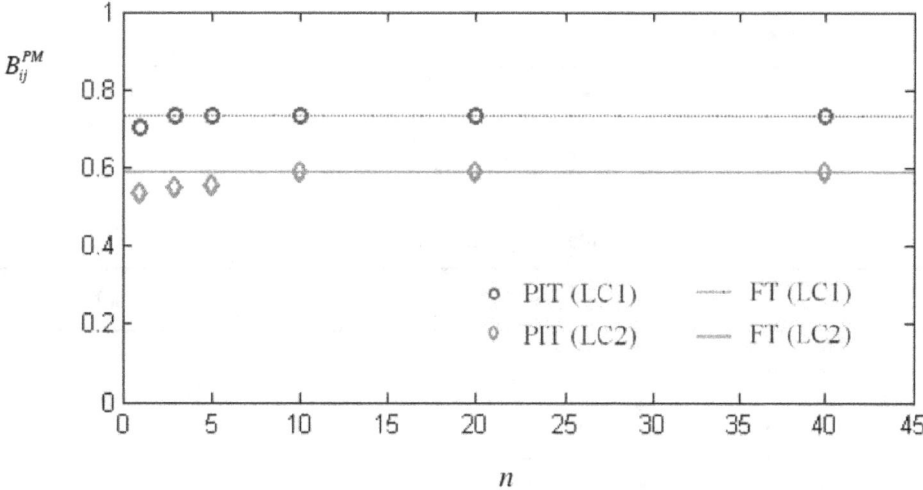

Figure 10. Estimation of DCIs from MPIT and FT

Table 3. Comparison of MPIT- and FT-based DCIs

	n	1	3	5	10	20	40
B_{ij}^{PM*}	N_m (out of 96)	14	5	5	2	1	1
	R_m	0.91	0.98	0.98	0.99	0.99	0.99
B_{ij}^{VT*}	N_m (out of 96)	30	7	3	3	3	3
	R_m	0.83	0.99	0.99	0.99	0.99	0.99

Table 4. Adjusted peak demand-to-capacity indexes

	Corner column	Non-corner column	Exterior beam	Interior beam
B_{ij}^{PM*}	0.88	0.89	0.56	0.56
B_{ij}^{VT*}	0.57	0.45	0.48	0.62

Table 5. Peak interstory drifts and peak acceleration

	x dir.	y dir.	resultant
Interstory drift ratio [$\times 10^{-4}$]	16	29	31
Acceleration [mg]	12.4	17.6	19.3

5. Concluding remarks

The DAD methodology using the multiple points-in-time (MPIT) approach was developed in this study to achieve the reliable and efficient estimation of peak combined wind effects. Instead of observing the peak wind effects from calculations based on all data points in a time history, the MPIT approach is based on a limited number of data points selected from peaks in the time series of individual effects.

The validity of the MPIT approach was investigated for a 60-story reinforced concrete building, known as the CAARC building. To obtain peak demand-to-capacity indexes (DCIs) (i.e., combined wind effects on structural members due to interacting forces and moments), various numbers n of peaks of the individual time series of each force and moment (i.e., $n = 1, 3, 5, 10, 20$, and 40) were used, and the DCIs were calculated at the points in time corresponding to those peaks. The highest DCI was selected as largest of these DCIs. The MPIT-based DCIs were compared with peaks of the full time-series, that is, with the peak DCIs for all data points in the time history. The comparisons showed that the MPIT approach based on $n \geq 3$ yielded reliable peak DCIs for all 96 structural members considered in the study.

The MPIT-based DAD developed in this study not only provides accurate combined wind effects not obtainable by the frequency domain approach commonly used in wind engineering practice, but also reduces significantly the amount of computational time required by a conventional time domain analysis using full time histories. The MPIT approach used in conjunction with DAD is therefore a practical and efficient design tool.

References

ACI (2008). *Building code requirements for structural concrete (ACI 318-08) and commentary*, American Concrete Institute, Farmington Hills, MI.

ASCE (2010). *Minimum design loads for buildings and other structures*, American Society of Civil Engineers, Reston, VA.

Ferry-Borges, J., and Castanheta, M. (1971). *Structural safety*, 2nd ed., National Civil Engineering Laboratory, Lisbon, Portugal.

Isyumov, N., Fediw, A. A., Colaco, J., and Banavalkar, P. V. (1992). "Performance of a tall building under wind action." *Journal of Wind Engineering and Industrial Aerodynamics*, 42(1-3), 1053-1064.

Kareem, A., Kijewski, T., and Tamura, Y. (1999). "Mitigation of motions of tall buildings with specific examples of recent applications." *Wind and Structures*, 2(3), 201-251.

Main, J. A., and Fritz, W. P. (2006). *Database-Assisted Design for Wind: Concepts, Software, and Examples for Rigid and Flexible Buildings*. NIST Building Science Series 180, National Institute of Standards and Technology, Gaithersburg, MD.

Melbourne, W. H. (1980). "Comparison of measurements on the CAARC standard tall building model in simulated model wind flows." *Journal of Wind Engineering and Industrial Aerodynamics*, 6(1-2), 73-88.

PCA (2008). *PCA notes on 318-08 building code requirements for structural concrete with design applications*, ed., Portland Cement Association, Skokie, IL.

Simiu, E. (2011). *Designing buildings for wind: a guide for ASCE 7-10 Standard users and designers of special structures*, 2nd ed., John Wiley & Sons, Hoboken, NJ.

Teshigawara, M. (2001). "Structural design principles (chapter 6)." in *Design of modern highrise reinforced concrete structures*, H. Aoyama, ed., Imperial College Press, London.

Turkstra, C. J., and Madsen, H. O. (1980). "Load combination in condified structural design." *Journal of the Structural Division-ASCE*, 106(12), 2527-2543.

Venanzi, I. (2005). *Analysis of the torsional response of wind-excited high-rise building*, Ph.D. Dissertation, Università degli Studi di Perugia, Perugia.

Wardlaw, R. L., and Moss, G. F. (1971). "A standard tall building model for the comparison of simulated natural winds in wind tunnels." *International conference on wind effects on buildings and structures*, Tokyo, Japan, 1245-1250.

Wen, Y. K. (1977). "Statistical Combination of Extreme Loads." *Journal of the Structural Division-ASCE*, 103(5), 1079-1093.

Yeo, D. (2010). *Database-assisted design for wind: Concepts, software, and example for of high-rise reinforced concrete structures*. NIST Technical Note 1665, National Institute of Standards and Technology, Gaithersburg, MD.

Yeo, D., and Simiu, E. (in press). "High-rise reinforced concrete structures: database-assisted design for wind." *Journal of Structural Engineering*.

www.ingramcontent.com/pod-product-compliance
Lightning Source LLC
Chambersburg PA
CBHW081814170526
45167CB00008B/3435